FLORA OF TROPICAL EAST AFRICA

POTAMOGETONACEAE

J.J. SYMOENS[1]

Aquatic herbs, perennial or rarely annual, glabrous, rooted in the substrate. Vegetative shoot system comprising horizontal shoots, mostly stoloniferous, non chlorophyllous (here called rhizomes) and vertical shoots, erect or sometimes ± floating, often densely branched, chlorophyllous (here called stems); tubers or turions sometimes present on the horizontal and/or the erect shoots. Leaves alternate or subopposite, rarely ternate, sessile or petiolate, simple, served by 3 traces, parallel-veined or with a single midvein; stipules present throughout or only on young shoots (*Groenlandia*), often forming a tubular sheath around the stem and young inflorescence, either free or adnate to the base of the blade; intravaginal scales 2–several, situated in leaf axils, mostly linear to subulate, rarely ovate, entire and acute; blade entire, denticulate or serrate. Inflorescences pedunculate, with 2 opposite flowers (*Groenlandia*) or more than 2 flowers in a dense or interrupted spike (*Potamogeton*); peduncle rigid and mostly slightly elevated above the water level, but sometimes flexuous, the flowers resting then on the water surface, rarely submerged, but often withdrawing the fruits below the surface; bracts abortive or absent. Flowers ⚥ (hermaphrodite), small, regular, hypogynous, tetramerous. Perianth: a single whorl of 4 valvate shortly clawed tepals. Stamens 4, opposite the tepals and basally adnate to the claw; anthers sessile, bilocular, extrorse, opening by longitudinal slits. Gynoecium superior, of (1–3) 4 (5–8) carpels, sessile, free or shortly connate at the base, alternating with the stamens; styles usually short; stigma unicellular-papillate and strongly secretory or smooth and dry or weakly secretory; ovule solitary in the carpel and attached to its ventral margin, orthotropous at first, but becoming campylotropous at maturity, pendant and filling the locule, bitegmic, crassinucellate. Fruit apocarpous consisting of distinct sessile fruitlets, ± drupaceous, but opening with a dorsal lid, with a rather fleshy parenchymatous mesocarp and a sclerified endocarp, the latter multi-layered (*Potamogeton*) or 1-layered (*Groenlandia*). Seed without endosperm; embryo unciform or spiral (i.e. coiled more than 1 complete turn), with a large hypocotyl and a single obliquely terminal cotyledon that encloses the plumule.

A family of 2 genera, *Potamogeton* L. with 69 species and cosmopolitan, with highest diversity in western Europe, Japan and eastern North America, and *Groenlandia* J.Gay, the latter with one species and restricted to Europe, western Asia and North Africa. A third genus has been created by elevation of *Potamogeton* subgen. *Coleogeton* Reichenb. to generic status under the name *Stuckenia* Börner (in Abh. Naturwiss. Ver. Bremen 21: 258, 1912) or *Coleogeton* (Reichenb.) Les & Haynes (in Novon 6: 389, 1996), but this view is contested by Teryokhin & Chubarov (in Bot. Zhurn. 81, 7: 23–33, 1996) and by Wiegleb & Kaplan (in Folia Geobot. 33: 241–316, 1998) and is not followed in this treatment.

[1] Laboratory for general Botany and Nature Management (APNA), Free University of Brussels, Pleinlaan 2, B-1050 Brussels, Belgium

POTAMOGETON

L., Sp. Pl.: 126 (1753); A. Bennett in F.T.A. 8: 219 (1901); Dandy in J.L.S. 50: 507–540 (1937); Wiegleb & Kaplan in Folia Geobot. 33: 241–316 (1998); Cook, Aquat. Wetl. Pl. S. Afr.: 234–239 (2004)

Rhizome present or absent, if present well differentiated from the stem, filiform, slender or robust, terete or ± compressed; stems terete or compressed, ± densely branched; turions present or absent. Leaves alternate, but mostly subopposite toward the inflorescence apex (involucral leaves), sessile or petiolate, 1- to many-nerved, all submerged and with filiform, grass-like or ± expanded blade, or dimorphic, the submerged ones thin and ± translucent, sometimes reduced to phyllodes, the floating ones ± coriaceous, opaque, petiolate, and usually broader; stipules always present, although sometimes fugacious, forming a sheath, connate or convolute around the stem; blade or petiole sometimes attached near the top of the sheath which projects beyond the joint as a ventral ligule, sometimes attached farther down the sheath or arising from the node. Spikes cylindrical or subglobose, 3- to many-flowered, dense or lax, sometimes interrupted, mostly borne above the water and wind-pollinated, but sometimes submerged and water-pollinated. Tepals rounded at the apex. Fruitlets with soft mesocarp and multi-layered sclerified endocarp, with or without a distinct beak. Embryo unciform or spiral; cotyledon curved in the seed.

A cosmopolitan genus with 69 recognised species and 50 recognised hybrids (according to the account by Wiegleb & Kaplan in Folia Geobot. 33: 241–316, 1998).

The identification of *Potamogeton* specimens may be complicated by the great variation in leaf form with age, or under influence of environmental factors like current speed, water depth, light intensity, and nutrient supply. The existence of numerous hybrids may also be troublesome. A binocular microscope should be used for the examination of the leaf veins and lacunae and of the fruitlet characters. As far as possible, one should always attempt to collect well developed, flowering and fruiting specimens.

Useful taxonomic characters for the identication may also be observed in the stem anatomy pattern (Fig. 1). This requires the preparation of thin cross-sections of the stem, preferably of the internode of the upper part of the flowering stem, for examination with a compound microscope. The main characters to be observed are the shape and size of the stele, the shape of the endodermis cells (mostly U or O), the number and size of interlacunar and of subepidermal bundles, and the presence or absence of pseudohypodermis.

Many *Potamogeton* species produce more or less modified buds (generally called turions, winter-buds or hibernacula) serving as means of vegetative spread or of perennation during unfavourable periods. Different types of such organs have been recognised in *Potamogeton* and provide valuable taxonomic criteria. They are produced from the rhizomes or are borne on the stems and their branches, either at their apex or in the axils of the leaves. Mostly, the leaves rather then the axis form their major components, but in some species, swollen internodes of the rhizomes form tubers, abundantly filled with starch and topped by a bud which will become the future erect stem. These structures, common in the temperate regions, are less well known in the African *Potamogeton* populations. More information is needed on their presence and their role in the life cycle of the tropical species.

FIG. 1. *POTAMOGETON RICHARDII* — stem anatomy in transverse section. **1**, transverse section of internode, diagrammatic, × 25: small dots indicate subepidermal fibrous bundles and remnants of former subepidermal bundles now deeper in lacunar cortex; large dots indicate interlacunar vascular bundles; the stele (or central cylinder) is here of trio type, with 10 vascular bundles, irregularly surrounded by sclerenchyma (thick lines); **2**, detail of peripheral tissues, × 200: ep, epidermis; ps, pseudohypodermis (here 1-layered); sub b, subepidermal bundle; **3**, detail of the contact zone of stele and cortex, × 200, from position indicated in 1: st lac, stelar xylem lacuna; scl, sclerenchyma; end, endoderm (here consisting of cells of U type); c par, cortical parenchyma (compact in the innermost layers); **4**, cortical lacunar system, × 200; **5**, detail of an interlacunar vascular bundle, × 200. All figures are oriented as in 1. From *Bamps, Symoens & Vanden Berghen* 526 (Zimbabwe). Drawn by J.J. Symoens.

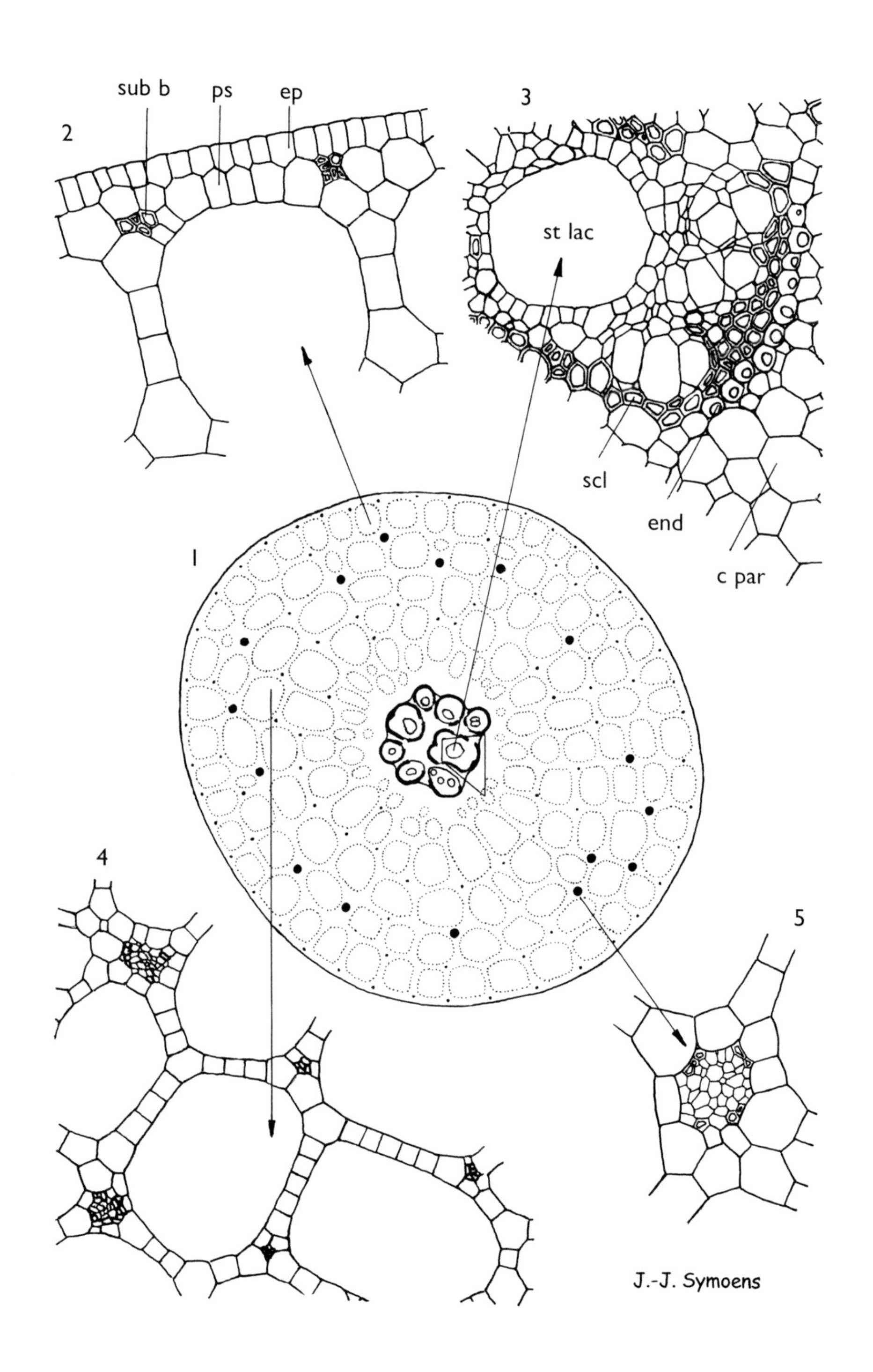
sub b
ps
ep
2
3
st lac
scl
end
c par
1
4
5
J.-J. Symoens

1. Floating leaves absent 2
   Floating leaves present, different from the submerged leaves, sometimes small 7
2. Leaves more than 5 mm broad 3
   Leaves less than 4 mm broad, narrowly linear to filiform 4
3. Submerged leaves sessile or very shortly petiolate (petiole mostly less than 30 mm), finely acute or mucronate, persistent; cells of the stem endodermis of U-type, interlacunar bundles present, pseudohypodermis present, 1-layered . . . . . 5. *P. schweinfurthii*
   Submerged leaves distinctly petiolate (petiole mostly 30–150 mm long), narrowly obtuse to subacute but never mucronate, often absent after fructification; cells of the stem endodermis of O-type, interlacunar bundles absent, pseudohypodermis absent or present as 1, often discontinuous layer . . . . . 6. *P. nodosus*
4. Stipules adnate to the blade, forming a winged leaf base and ending in a free, hyaline ligule, later disintegrating in fibres . . . . . 1. *P. pectinatus*
   Stipules present as a sheath round the stem and/or axillary bud 5
5. Carpels 1–2 in each flower; leaves usually 0.3–1 mm broad, lateral veins absent or indistinct, lacunae absent or restricted to a very narrow band at the base of the leaf, rarely extending almost to the apex; fruitlet dorsal edge often muriculate . . . . . 3. *P. trichoides*
   Carpels 3–7 in each flower; leaves usually 1–2 mm broad, with or without a row of lacunae bordering the midrib 6
6. Leaves with 1–2 distinct veins on each side of midrib; lacunar system bordering the midrib absent or narrow; fruitlets 1.8–2.7 mm long, dorsally without keel and smooth . . . . . 2. *P. pusillus*
   Leaves with 1 inconspicuous vein on each side of midrib; lacunar system bordering the midrib very broad; fruitlets 1.5–2.4(–3.5) mm long, with dorsal keel mostly distinct and crenulate . . . . . 4. *P. octandrus*
7. Submerged leaves less than 3 mm broad; floating leaves with petiole 3–25(–34) mm long and lamina (5–)9–29 (–38) mm long; spikes 5–16 mm long in fruit . . . . . 4. *P. octandrus*
   Submerged leaves, if present, more than 3 mm broad, the lower ones often reduced to phyllodes; floating leaves with petiole (14–)30–210 mm long and lamina 30–150 mm long; spikes 25–90 mm long in fruit 8
8. Submerged leaves disappearing early and not present in the mature plant; floating leaves with petiole 18–110 (–200) mm long, often with a discoloured section at the junction of the lamina; cells of the stem endodermis of U type; interlacunar bundles present in (1–)2–3 circles, pseudohypodermis present, 1-layered . . . . . 7. *P. richardii*
   Submerged leaves present and mostly persistent; floating leaves with petiole (14–)30–70(–210) mm long, without discoloured section at the junction of the lamina 9

9. Submerged leaves sessile or very shortly petiolate (petiole mostly less than 30 mm long, but up to 82 mm in intermediate leaves); blade 45–190(–260) mm long, (3–)7–28 mm broad, 4–21 times as long as broad, with apex finely acute to sharply mucronate; floating leaves with petiole (14–)30–70(–200) mm long; spikes 30–90 mm long in fruit; cells of the stem endodermis of U-type; interlacunar bundles present, in 1(–2) circles; pseudohypodermis present, 1-layered .... 5. *P. schweinfurthii*

Submerged leaves petiolate (petiole 20–150 mm), 30–190 (–280) mm long, 10–38(–50) mm broad, 5–9 times as long as broad, with apex narrowly obtuse to subacute, but never mucronate; floating leaves with petiole 30–210 mm long; spikes 14–50 mm long in fruit; cells of the stem endodermis of the O-type, interlacunar bundles absent, pseudohypodermis absent or present as 1, often discontinuous, layer .... 6. *P. nodosus*

1. **Potamogeton pectinatus** *L.*, Sp. Pl.: 127 (1753); K. Schum. in Engl., P.O.A. C: 93 (1895); A. Benn. in F.T.A. 8: 223 (1901) (excl. *G. Don*); A. Benn. in J.L.S. 38: 23 (1907); Graebner in Z.A.E. 2: 42 (1910); Hagstr. in R.E. Fries, Wiss. Ergebn. Schwed.-Rhod.-Kongo-Exped.: 188 (1916); A. Peter in Abh. Ges. Wiss. Göttingen 13 (2): 18, 108 (1928), & in F.D.O.-A.: 113 (1929); Dandy in J.L.S. 50: 513, map fig. 1 (1937); Robyns & Tournay, F.P.N.A. 3: 14 (1955); Lind & Tallantire, F.P.U., ed 1: 194 (1962) & ed. 2: 194 (1972); Lisowski *et al.* in F.A.C., Potamogetonaceae: 3 (1978); U.K.W.F., ed. 2: 302, t. 134 (1994); Lye in Fl. Eth. & Eritr. 6: 19, fig. 176.1.1 (1997); Wiegleb & Kaplan in Folia Geobot. 33: 305 (1998); Cook, Aquat. Wetl. Pl. S. Afr.: 236, fig. 254 a-c (2004). Type: Austria, *Burser* s.n. (UPS, Burser Herb. X:124, lecto., chosen by Haynes in Taxon 35: 569 (1986))

Rooted submerged aquatic with adventitious roots at the nodes of the horizontal shoots; rhizomes perennial, slender to very robust, terete, whitish, sometimes developing tubers filled with starch and surrounded by a scaly leaf at the end of the growing season (such tubers generally absent in tropical plants); stems annual to perennial, up to 4 m long and 0.5–2 mm in diameter, filiform to relatively robust, terete, sometimes pink, usually richly branched. Submerged leaves bright green to olive green, sessile, (20–)30–125(–300) mm long, 0.2–4 mm wide, 24–200 times as long as wide, linear, canaliculate or flattened, straight at base, entire at margins, the narrower leaves acute to finely acuminate at apex, sometimes the broader leaves obtuse and mucronate; midrib bordered on each side by one to several air channels, lateral veins 1–2 on each side, inconspicuous; stipules persistent, fused with the leaves for 8–70 mm, forming a convolute and winged leaf base, ending in a free hyaline ligule, (3–)5–15 mm long, obtuse, rounded or truncate, later disintegrating into fibres; intravaginal scales 0.4–0.6 mm long; floating leaves always absent. Peduncles 20–100(–450) mm long, 1.5–5(–10) times as long as the fruiting spike, as thick as the stem, flexuous; spikes (4–)8–14-flowered, with 2–7 pairs or whorls of flowers, contiguous at first, later distant, or the upper groups ± contiguous, 13–35(–60) mm long when in fruit. Tepals orbicular to elliptical, 1–3 mm long; anthers 0.8–1.3 mm long; carpels 4; stigmas borne on a short but distinct style. Fruitlets brown, (2.5–)3.3–4.7(–5.1) mm long, 2–3.6 mm broad, asymmetrically obovoid, ventrally nearly straight, dorsally very convex, hardly 3-keeled; beak 0.2–0.6 mm long, ventral, rarely subventral. Stem anatomy: stele of four bundles, mostly oblong; endodermis of U-type; interlacunar bundles present (in 1 ± complete ring); subepidermal bundles few or absent; pseudohypodermis present, 1-(–2)-layered. Fig. 2 (p. 6).

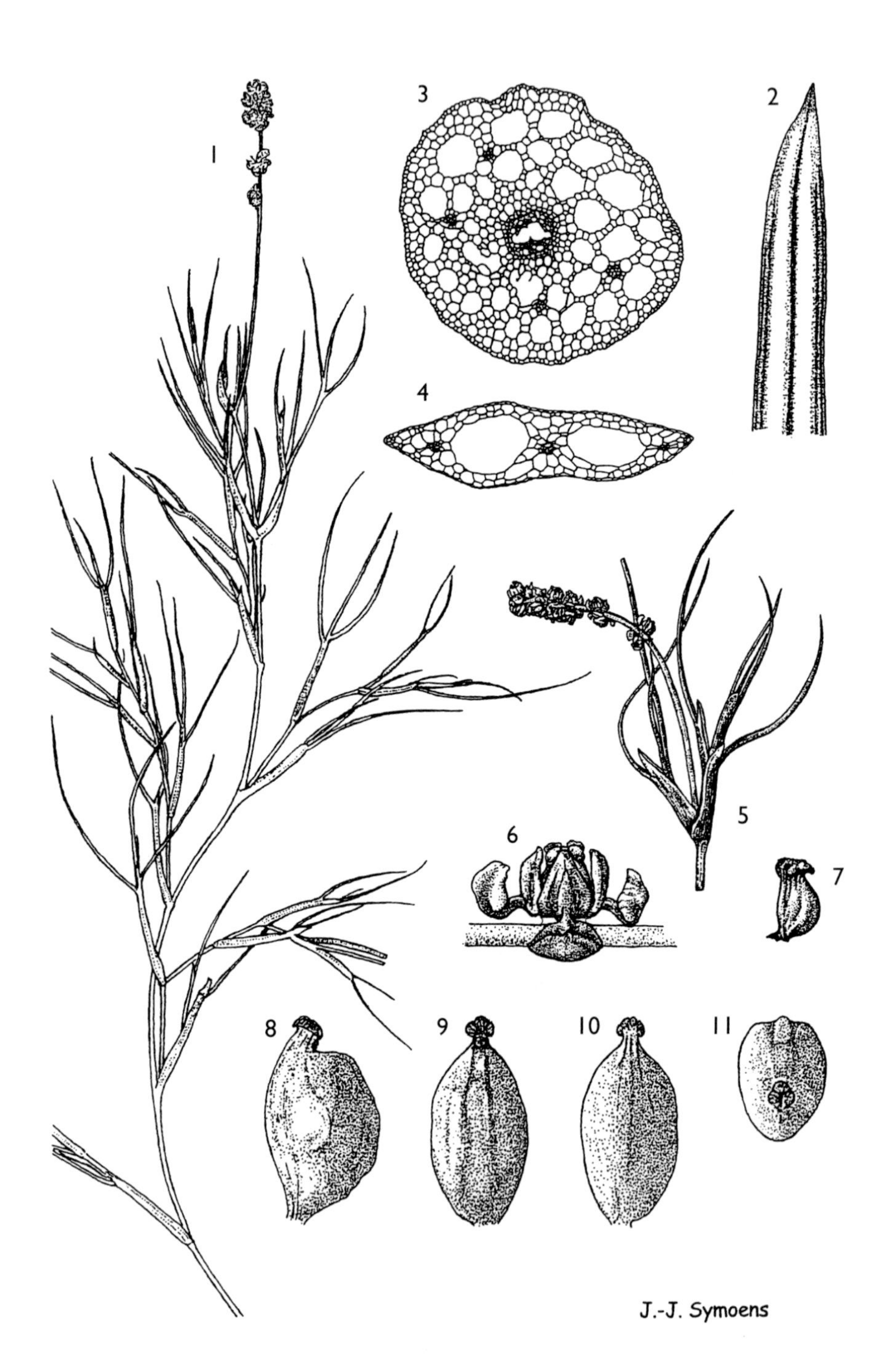

FIG. 2. *POTAMOGETON PECTINATUS* — **1**, habit, × 1 ; **2**, leaf tip, × 20; **3**, T.S. of stem, × 40; **4**, T.S. of leaf upper part, × 40; **5**, node and inflorescence, × 1.2; **6**, open flower, tepals spread, × 8; **7**, carpel, × 8; **8–11**, fruitlet, respectively lateral, dorsal, ventral and apical view, × 8. 1 from *LaBarbara* 16; 2–4 from *Ross* 1469; 5–7 from *Symoens* 5340; 8–11 from *Lye* 5329. Drawn by J.J. Symoens.

UGANDA. Kigezi District: Lake Bunyonyi, Apr. 1945, *Purseglove* 1644!; Toro District: Nyakasura, near Fort Portal, crater lake, 4 Dec. 1934, *G. Taylor* 2336!; Bunyoro District: near Butiaba, Lake Albert, 8 Feb. 1912, *Fries* 2000!

KENYA. Northern Frontier District: Lake Turkana, SE bank, 20 March 1970, *Bally* s.n.!; Turkana District: Lake Turkana, Ferguson's Gulf, 4 June 1951, *McKay* s.n.! Naivasha District: Lake Naivasha, Apr. 1938, *Chandler* 2270!

TANZANIA. Musoma District: Upper Mugungu [Mgungu], 23 Oct. 1961, *Greenway, Myles Turner & Harvey* 10284!; Masai District: Ngorongoro Crater, Gaitokitoke Springs, 1956, *Greenway* 9063!; Kigoma District: Kigoma, Lake Tanganyika, 31 Jan. 1953, *Ross* 1469!

DISTR. **U** 2; **K** 1–4; **T** 1, 2, 4, 5, 7; subcosmopolitan, widely distributed in Africa, including North Africa, Madagascar and the Mascarenes Islands

HAB. Lakes, rivers, irrigation canals, also in brackish and in polluted water; water depth up to 3 m; forming often extensive monospecific patches; also a serious water weed, often blocking the waterflow in irrigation and drainage canals; 350–2300 m

SYN. *Potamogeton livingstonei* A. Benn. in F.T.A. 8: 223 (1901); Graebner in Pflanzenr. 4 (11): 128 (1907). Types: Malawi, Lake Malawi, Kambwe Lagoon, 1877, *Laws* 3 (BM!, lecto., chosen here, K!, isolecto.)

[*P. filiformis* sensu A. Benn. in F.T.A. 8: 223 (1901); Graebner in Pflanzenr. 4 (11): 126 (1907) quoad *Scott Eliott* 3379b; A. Peter in Abh. Ges. Wiss. Göttingen 13 (2): 18, 107 (1928); F.D.O.-A.: 113 (1929), *non* Pers.]

Stuckenia pectinata (L.) Börner, Fl. Deutsch. Volk: 713 (1912)

Coleogeton pectinatus (L.) Les & Haynes in Novon 6: 390 (1996)

NOTE. *Potamogeton pectinatus* is an extremely polymorphic species, with many regional and local forms. Many subspecific taxa were described, mostly at the variety level, but there are many transition forms and transplantation experiments have showed that characters used to distinguish the infraspecific taxa may be environmentally induced. The very robust specimens from the African Great Rift lakes, with 2–4 mm wide leaves (e.g. *Worthington* 230–3), were originally described as *P. livingstonei.*

P. pectinatus differs from *P. filiformis* Pers. in having open stipular sheaths, larger fruits and a short but distinct style.

An important food source for many waterfowl species.

2. **Potamogeton pusillus** *L.*, Sp. Pl.: 127 (1753); A. Benn. in F.T.A. 8: 222 (1901); U.K.W.F., ed. 2: 303 (1994); Lye in Fl. Somal. 4: 14, fig. 8A, B (1995) & in Fl. Eth. & Eritr. 6: 20, fig. 176.2.1–3 (1997); Wiegleb & Kaplan in Folia Geobot. 33: 292 (1998); Cook, Aquat. Wetl. Pl. S. Afr.: 237, fig. 254 d–g (2004). Type: Europe (LINN 175.15, left hand specimen, lecto., chosen by Dandy & Taylor in J.B. 76 : 92 (1938))

Submerged aquatic herb; rhizomes absent or only present later in the growing season, annual to biennial, filiform, terete, with short internodes; stems annual to perennial, up to 1 m long, 0.3–0.7(–1) mm in diameter, terete or compressed, sparingly to much branched; turions, when present, mostly sessile and axillary, but sometimes terminal on axillary branches, rigid, fusiform, with a few erect, patent or recurved free leaves. Submerged leaves bright green to olive green or dark green, sometimes with a brownish tinge, sessile, linear, (9–)20–85(–110) mm long, (0.3–) 0.8–2(–2.5) mm wide, (15–)20–90 times as long as wide, flaccid or firm, translucent, narrowly cuneate at base, tapering or rather abruptly narrowed to an acute or acuminate apex; margins entire, bordered by a narrow marginal vein; midrib occupying 15–35 per cent of the leaf width at base, not bordered by lacunae or the lacunae poorly developed and restricted to the lower leaf half; lateral veins 1(–2) on each side, distinct, joining the midrib 1.5–4 leaf widths below the leaf apex; true floating leaves absent but rarely the uppermost leaves with lamina floating at the water surface, subsessile, linear-oblanceolate, 18–38 mm long, 1.3–3.1 mm wide, 7–20 times as long as wide, bright green, narrowly cuneate at base, acute to narrowly obtuse at apex, 3–5-veined, with broad rows of lacunae bordering the midrib; stipules axillary, connate and forming a tubular sheath at least in their lower part and when young, but splitting with age, 4–18(–32) mm long, translucent, persistent or decaying; turions mostly sessile and axillary, but sometimes also terminal on axillary branches, narrowly cylindrical, with a few erect, patent or recurved free leaves. Peduncles filiform to slightly clavate, ± flexuous, (6–)10–30(–55) mm long, 1–6 times as long as the fruiting spike, as thick as the stem, slightly or distinctly compressed; spikes cylindrical, with 2–7 flowers, in 1–4 whorls, ± contiguous, sometimes the lowest flower remote, 4–15 mm long in fruit. Tepals 0.8–1.8 mm long, mostly persistent; carpels (3–)4(–5); anthers 0.7–0.9 mm long. Fruitlets obovoid, ± compressed, 1.8–2.7 mm long, 1–1.5 mm broad, green to pale olive, dorsally without keel, beak nearly centrally placed, straight or somewhat oblique, 0.2–0.4 mm long. Stem anatomy: stele of the circular type; endodermis of the O-type; interlacunar bundles absent; subepidermal bundles present; pseudo-hypodermis mostly absent, if present 1-layered.

UGANDA. Mengo District: Entebbe, Lake Victoria, Nov. 1930, *Snowden* 1836!
KENYA. Laikipia District: Ngobit [Ongobit], Sharpe's Farm, July 1954, *Bally* B9775!; Kiambu/Machakos Districts: Athi R., Fourteen Falls, 1956, *Verdcourt* 1571!
TANZANIA. Rungwe District: 8 km from Kyela, Nov. 1969, *Wingfield* 482!
DISTR. **U** 4; **K** 3, 4; **T** 7; Mauritania, Sudan, Ethiopia, Eritrea, Somalia, Congo Kinshasa, Zambia, Malawi, Zimbabwe, Namibia, Botswana, Swaziland, Lesotho, South Africa; from Morocco to Libya and Egypt, also common in Europe and temperate parts of Asia, Arabia and North America, rarer in American and Asian tropics
HAB. In lakes, dams and ponds, also in slow-flowing waters, sometimes in slightly brackish water; water depth 30–60 cm; 600–1700 m

SYN. *P. panormitanus* Biv. in A. Biv. f., Nuov. Piant. ined.: 6 (1838); Dandy in J.L.S. 50: 523, map fig. 5 (1937); Lisowski *et al.* in F.A.C., Potamogetonaceae: 4 (1978). Type: Italy, Sicily, Palermo, Oreto R. & Gurgo di Rebuttone, no collector indicated (?FI, not located)
[*P. friesii* sensu A.Benn. in F.C. 7: 48 (1897); Medley Wood, Handb. Fl. Natal: 143 (1907), *non* Ruprecht]
[*P. preussii* A.Benn. in F.T.A. 8 (1901) pro parte, quoad syntypos sequentes: Eritrea, near Saganeita, Degerra Gorge, *Schweinfurth & Riva* 896 (BM!, BR!, FT!, K!, LD!); Ethiopia, Shire [Shireh], *Quartin-Dillon & Petit* 571 (K); Tigray, near Adwa, *Schimper* 179 (BM!, BR!, FI-W!, K!, L!); Lake Abbé [Amba Sea], *Schimper* 570 (K!)]
P. pusillus L. var. *africanus* A.Benn. in Ann. Cons. Jard. Genève 9: 102 (1905); Graebner in Pflanzenr. 4 (11): 115 (1907); Graebner in Z.A.E. 2: 42 (1910). Syntypes: South Africa, Transvaal, *Wilms* 1656, 1657 (G!); Natal, Durban (Port Natal), between Tyger Berg and Blue Berg, *Drège* 1206 (G!; iso at BM!, K!, S) & Umlazi Riv., *Drège* 4458 (G!); Natal, between Umzinto and Ifafa, *Wood* 3055 (BOL, G!, K!, NH)

NOTE. *P. pusillus* is a very polymorphic species, varying greatly in leaf length, the number of leaf nerves and rows of lacunae along the midrib, the shape of leaf apex, and the number of flower whorls in the inflorescence. Due to the lack of a consistently correlated set of morphological characters, no delimitation of infraspecific taxa based on these criteria seems possible. The connate stipules forming a tubular sheath, at least when young, differentiate *P. pusillus* from *P. berchtoldii* Fieb., a taxon of the temperate northern hemisphere, having stipules open throughout their length; isozyme studies confirmed the separate position of *P. pusillus* and *P. berchtoldiii* [Hettiarachchi & Triest in Opera Bot. Belg. 4: 87–114 (1991), Kaplan & Stepánek in Plant Syst. Evol. 239: 95–112 (2003)]. For the differences between *P. pusillus* and *P. trichoides*, see the note under *P. trichoides*.

3. **Potamogeton trichoides** *Chamisso & Schltd.* in Linnaea, 2 (2): 175, t. 4, fig. 6 (1827); Dandy in J.L.S. 50: 520, distr. map fig. 3 (1937); U.K.W.F., ed. 2: 302, t. 134 (1994); Wiegleb & Kaplan in Folia Geobot. 33: 297 (1998); Cook, Aquat. Wetl. Pl. S. Afr.: 238, fig. 256 c–f (2004). Type: Europe (lectotypification needed)

Submerged aquatic herb; rhizomes absent or only present at the end of the growing season, filiform, terete, sometimes developing terminal or axillary tubers; stems annual to perennial, up to 1.5 m long, sparingly to much branched, filiform, terete to slightly compressed, fragile and breaking easily at the nodes; turions apical or axillary, not greatly differing from the ordinary buds, but formed of more crowded leaves, darker and more rigid, detaching from the stem of the parent plant and easily dispersed by water (in the tropics ?). Submerged leaves bright green to dark green, often with a brownish tinge, sessile, linear, 14–80(–130) mm long, 0.3–1(–1.8) mm wide, 30–110 times as long as wide, narrowly cuneate at base, acuminate at apex, entire, 3-veined; midrib prominent, occupying 30–70 per cent of the leaf width at the base, not bordered by rows of lacunae or lacunae poorly developed and confined to the basal half of the leaf; lateral veins sometimes inconspicuous, without additional sclerenchymatous strands; secondary and marginal veins absent; floating leaves absent; stipules axillary, open and tightly inrolled, 5–27 mm long, translucent, often with a greenish tinge, obtuse at apex, persistent. Peduncles 10–75 mm long, (2–)3–9 times as long as the fruiting spike, as thick as the stem; spikes shortly cylindrical, 3–9 mm long when in fruit, with 3–5 flowers, contiguous to shortly distant near the base. Flowers 1(–2)-carpellate. Fruitlets 2.5–3.2 mm long; dorsal keel mostly distinct, often muriculate; beak 0.3–0.5 mm long, straight. Stem anatomy: stele of the circular type; endodermis of the O-type; interlacunar bundles absent; subepidermal bundles present; pseudohypodermis absent.

UGANDA. Kigezi District: Lake Bunyonyi, Chabahinga, Dec. 1938, *Chandler & Hancock* 2721! & Lake Mutanda, 31 Jan. 1939, *Loveridge* 454!; Ankole District: Lutoma Dam, Aug. 1953, *Lowe* 665!
KENYA. Nairobi District: Ngong Forest, Karen Pond, 4 Nov. 1934, *G. Taylor* 1594a!; Kiambu/Machakos Districts: Athi R., Fourteen Falls, 18 Apr. 1948, *Bogdan* 1564!
TANZANIA. Njombe District: Poroto Mts, Njombe, Nov. 1969, *Wingfield* 481!
DISTR. **U** 2, 4 (fide EA); **K** 4; **T** 7; Botswana, Zimbabwe and S Africa; Morocco, Algeria, Egypt, Europe and Asia
HAB. Dams, ponds and seasonal pools; water depth 30–50 cm; sometimes associated with *Potamogeton pusillus*; 1200–2450 m

NOTE. As mentioned by Wiegleb & Kaplan (*loc. cit.*, p. 298), *P. trichoides* is closely related to *P. pusillus* and it is sometimes difficult to distinguish specimens of *P. trichoides* in the vegetative stage from morphotypes of *P. pusillus* with extremely narrow leaves. *P. trichoides* differs from *P. pusillus* by having open and convolute stipules and a leaf midrib occupying 30–70 per cent of the leaf width at the base. When in flower or in fruit, *P. trichoides* may be distinguished by its gynoecium reduced to 1 or 2 carpels and its fruitlets with a distinct dorsal keel.

4. **Potamogeton octandrus** *Poir.* in Lam., Encycl. Méth. Bot., Suppl. 4: 534 (1816); Dandy in J.L.S. 50: 517, fig. 2 (1937); Hepper in F.W.T.A., ed. 2, 3: 16 (1968); Lisowski *et al.* in F.A.C., Potamogetonaceae: 5 (1978); U.K.W.F., ed. 2: 302 (1994); Cook, Aquat. Wetl. Pl. S. Afr.: 236, fig. 253 d–e (2004). Type: Viet-Nam [Cochinchina], *Loureiro* s.n. (BM!, holo. or lecto.)

Aquatic herb with submerged and floating leaves; rhizomes absent or present, annual or perennial, poorly developed, filiform, terete; stems annual, 20–100 cm long, sparingly or much branched, filiform, terete with grooves, sometimes developing axillary tubers. Submerged leaves bright green to brown-green, sessile, linear, 25–55(–75) mm long, 0.5–1.2(–3) mm wide, 30–75 times as long as wide, entire at margins, acute to acuminate, 3-veined; midrib bordered with broad rows of lacunae; lateral veins sometimes inconspicuous; intermediate leaves often present, petiolate, oblong to lanceolate; floating leaves present or absent, petiolate; petiole 3–25(–34) mm long, 0.2–1.1 times as long as the lamina; lamina narrowly lanceolate to elliptical or oblong, (5–)9–29(–38) mm long, (2–)3–11 mm wide, 1.3–6(–8.5) times as long as wide, opaque, coriaceous to subcoriaceous, bright green to dark green, sometimes with a brownish tinge, 5–7-veined, cuneate at base, acute at apex; margins entire; stipules axillary, convolute, 4–13 mm long, translucent, decaying early. Inflorescences developing mostly in the axils of floating leaves; peduncles 9–20(–40) mm long, 0.7–2.4 times as long as the fruiting spike, as thick or slightly thicker than the stem; spikes cylindrical, 5–16 mm long when in fruit, with 7–9 flowers in 3–7 contiguous or shortly distant whorls. Tepals broadly ovate, ± 1 mm long, green; carpels (3–)4(–7). Fruitlets 1.5–2.4(–3.5) mm long, green; beak short to fairly long, hooked; dorsal keel mostly distinct, crenulate. Stem anatomy: stele of the oblong or the circular type; endodermis of O-type; interlacunar bundles absent, rarely a few present; subepidermal bundles present; pseudohypodermis absent or present, 1-layered.

UGANDA. Busoga District: Buluba Leper Colony, 3 Oct. 1952, *G.H.S. Wood* 461!; Masaka District: Kagera River, near mouth, 9 Dec. 1951, *Norman* 63!; Mengo District: Entebbe, Lake Victoria, Jan. 1938, *Chandler* 2088!
KENYA. Uasin Gishu District: Eldoret, Lessos Dam, 18 Dec. 1964, *Lind, Agnew & Kettle* 5892!; Nairobi District: Nairobi National Park, Hyena Dam, 2 Jan. 1968, *Agnew* 9920!; Kiambu/Machakos Districts: Athi R., Fourteen Falls, 18 Apr. 1948, *Bogdan* 1565!
TANZANIA. Bukoba District: Bukoba, 20 Apr. 1905, *Cunnington* 54!; Iringa District: Mufindi, Lake Luisenga, 17 Mar. 1962, *Polhill & Paulo* 1788!
DISTR. **U** 1–4; **K** 3, 4; **T** 1, 4, 7; West Africa from Senegal to Nigeria, Cameroon, Chad, Central African Rep., Sudan, Ethiopia, Congo-Kinshasa, Burundi, Zambia, Malawi, Mozambique, Zimbabwe, Namibia, Botswana, South Africa, Madagascar; also in S and E Asia, and Australia
HAB. Lakes, ponds, backwaters, swamps, rice fields, also in swift flowing waters; water depth 0.6–2 m; associated with *Najas horrida, Potamogeton richardii, Ludwigia stolonifera, Utricularia inflexa*, etc.; 1100–1750 m (–2100 m, fide EA)

SYN. *Potamogeton javanicus* Hasskarl in Verh. Kon. Natuurk. Ver. Ned. Indie 1 (8): 26 (1856); K. Schum. in P.O.A. C: 93 (1895); A. Benn. in F.T.A. 8: 220 (1901) & in J.L.S. 38: 24, 28 (1907); Graebner in Pflanzenr. 4 (11): 46, fig. 14 A–C (1907); A. Peter in Abh. Ges. Wiss. Göttingen 13 (2): 18, 108 (1928); F.D.O.-A.: 113 (1929). Type: Indonesia: W Java, near Tjisarupan, foot of Mt. Papandayan, *Hasskarl* s.n. (not located)
P. tenuicaulis F.Muell., Fragm. Phyt. Austr. 1: 90 & 244 (1859) & 8: 217 (1874). Type: Australia, *F. Mueller* s.n. (MEL, holo.)
P. parvifolius Buchenau in Abh. Natur. Ver. Bremen 7: 32 (1880). Syntypes: Madagascar, Antananarivo, 17 Dec. 1877, *Rutenberg* s.n. (not located); India, Khasia Hills, *Hooker & Thomson* s.n. (K!, syn.)
P. preussii A.Benn. in F.T.A. 8: 222 (1901) pro parte quoad syntypes from Cameroon: Kumba, Barombi-ba-Mbu, Elephant Lake, 1890, *Preuss* 451 (BM, COI, syn.); Kumba Mountains [Johann-Albrechtshöhe], bank of Elephant Lake, 1895, *Staudt* 462 (BM, syn.)

NOTE. Lye (in Lidia 3: 79 (1993)) described as *Potamogeton octandrus* subsp. *ethiopicus* Lye a plant from Ethiopia with large fruits (3–3.5 mm long) and slender peduncles (0.3–0.5 mm thick).

5. **Potamogeton schweinfurthii** *A.Benn.* in F.T.A. 8: 220 (1901), pro parte quoad *Schimper* 1359, *nom. conserv.*, & in J.L.S. 38: 23, 28 (1907); Graebner in Pflanzenr. 4 (11): 79, fig. 19 (1907); Dandy in J.L.S. 50: 526, t. 21, 22 and map fig. 6 (1937); F.P.S. 3: 234 (1956); A.V.P.: 32 (1957); F.P.U., ed. 1: 194 (1962) & ed. 2: 194 (1972); Obermeyer in F.S.A. 1: 66, fig. 18 (2) (1966); Hepper in F.W.T.A., ed. 2, 3: 16 (1968); Lisowski *et al.* in F.A.C., Potamogetonaceae: 6, t. 1 (1978); Cribb & Leedal, Mountain Fl. S. Tanzania: 20 (1982); Iversen in Symb. Bot. Ups. 29 (3): 233 (1991);U.K.W.F, ed. 2: 302, t. 135 (1994); Lye in Fl. Somal. 4: 14, fig. 8C, D (1995) & in Fl. Eth.& Eritr. 6: 23, fig. 176.2.4–5 (1997); Wiegleb & Kaplan in Folia Geobot. 33: 273 (1998); Cook, Aquat. Wetl. Pl. S. Afr.: 237, fig. 255 (2004); Kaplan & Symoens in J.L.S. 148: 346, fig. 7, 8, fig. 9 map (2005). Type: Ethiopia, Lake Tana, *Schimper* 1359 (K!, lecto., chosen by Kaplan & Symoens in Taxon, 53: 837 (2004), isolecto. BM!, CGE, E, LD!)

Rooted aquatic herb, with submerged and floating leaves; rhizomes perennial, slender, terete, white, with apical winter buds; stems annual, sparingly to much branched, slender, terete, up to 3.5 m long, ± spongy; specialized dormant turions not developing. Submerged leaves almost always present, the young ones yellow-green or brownish, the old ones bright green or dark olive-green, often with a reddish or brownish tinge, sessile to shortly petiolate, often incrusted with calcium carbonate; petiole 0–30(–82) mm long; lamina narrowly lanceolate to oblong-elliptical, sometimes that of the lower leaves reduced to phyllodes, 45–190(–260) mm long, (3–)7–28 mm wide, 4–70 times as long as wide, membranous and translucent, cuneate, acute to mucronate, 5–11(–13)-veined, with or without narrow rows of lacunae bordering the midrib, margins often slightly undulate, entire or bearing very minute hyaline teeth; intermediate leaves sometimes present, instead of floating leaves; floating leaves absent or present, yellow-green to dark green, with a reddish tinge; petiole (14–)30–70(–200) mm long, never discoloured at junction

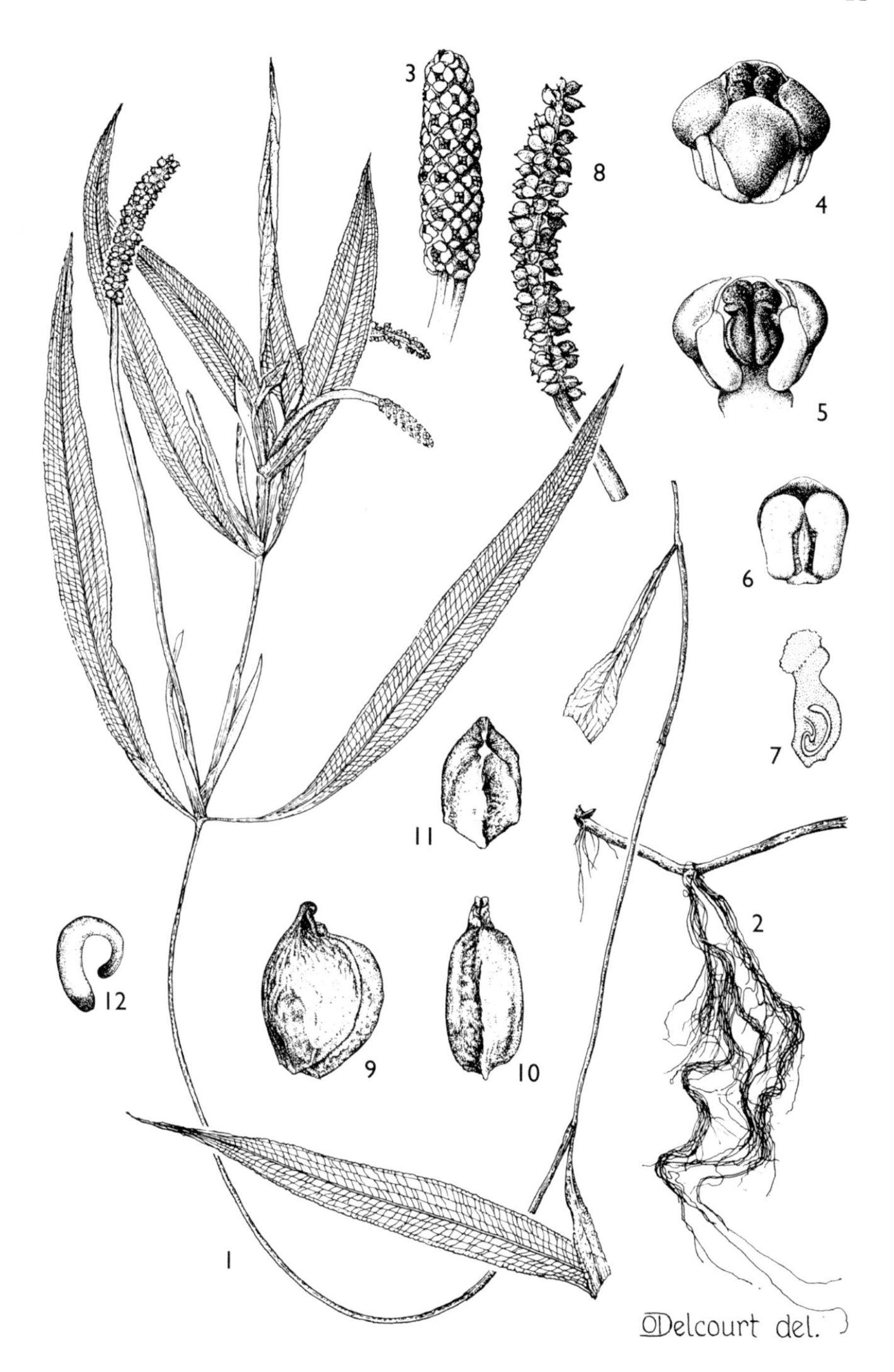

FIG. 3. *POTAMOGETON SCHWEINFURTHII* — **1**, habit, × 1/2; **2**, part of rhizome, × 1/2; **3**, flowering spike, × 1.3; **4**, flower, × 6.5; **5**, flower, one tepal and stamen removed, × 6.5; **6**, stamen and tepal, × 6.5; **7**, carpel L.S., × 10; **8**, fruiting spike, × 1.3; **9**, fruitlet, lateral view, × 6.5; **10**, fruitlet, dorsal view, × 6.5; **11**, fruitlet, apical view, × 6.5; **12**, seed, × 6.5.
1, 3–11 from *Symoens* 10609; 2 from *Hendrickx* 7992; 12 from *Hulot* 52. Drawn by O. Delcourt, and reproduced from F.A.C. with permission.

with the lamina; lamina narrowly oblong to elliptical or ovate, 43–130 mm long, 12–30 mm wide, 2–6 times as long as wide, opaque, subcoriaceous to coriaceous, 11–21-veined, entire, mostly narrowly cuneate at base, obtuse to acute; stipules axillary, convolute, 20–62 mm long, translucent, acute, persistent. Peduncles mostly terminal or lateral in the axils of submerged leaves, sometimes in the axils of floating leaves, 35–250 mm long, 2–3 times as long as the fruiting spike, thicker than the stem; spikes emergent, cylindrical, 30–90 mm long when in fruit, contiguous. Flowers numerous, sepals pale greeny brown; anthers yellow green; carpels 4; stigmas whitish becoming mauve-brown. Fruitlets pale brown-green, 2.9–3.9(–4.1) mm long, 2–2.5 mm wide, ventrally nearly straight; dorsal keel distinct, lateral keels sometimes present; beak short, obtuse. Stem anatomy: stele of the trio type, rarely of the proto or oblong type, endodermis of the U-type, interlacunar bundles present, in 1 circle (rarely 2, one being incomplete), subepidermal bundles absent or scattered ones present, pseudohypodermis present, 1-layered. Fig. 3 (p. 11).

UGANDA. Bunyoro District: junction of Victoria Nile and Lake Albert, 10 Feb. 1935, *Taylor* 3358a!; Mengo District: Entebbe, Jan. 1938, *Chandler* 2089! & Lake Victoria, Macdonald Bay, 14 Jan. 1953, *G.H.S. Wood* 574!

KENYA. Naivasha District: Naivasha, 13 Mar. 1936, *Meinertzhagen* EA 9468!; Kiambu District: Ngong Forest, Karen Pond, 4 Nov. 1934, *G. Taylor* 1594!; Kisumu-Londiani District: Dunga Beach, near Kisumu, Lake Victoria, 22 March 1996, *Symoens* 15012!

TANZANIA. Arusha District: Lake Duluti, 14 Nov. 1901, *Uhlig* 510!; Lushoto District: West Usambara Mts, km 1 on Shume–Lushoto road, 19 Nov. 1975, *Leeuwenberg* 10816!; Kigoma District: Kigoma, Lake Tanganyika, 31 Jan. 1953, *Ross* 1471!

DISTR. **U** 1–4; **K** 3–5; **T** 1–5; West Africa from Senegal to Niger, Central African Republic, Chad, Sudan, Ethiopia, Eritrea, Somalia, Congo (Kinshasa), Rwanda, Burundi, Zambia, Malawi, Mozambique, Zimbabwe, Namibia, Botswana, South Africa; Madagascar; from the Azores to Libya and Egypt and possibly the Middle East

HAB. Lakes, ponds, pools, backwaters, also in running waters; the leaves are sometimes incrusted with calcium carbonate; water depth 20–200 cm; associated with *Najas*, *Ceratophyllum*, *Nymphaea*, *Trapa*, etc.; 600–4250 m

SYN. ["?*P. lucens*" Thoms. in Speke, J. Discov. Source Nile: 651 (1863), *non* L.]

[*P. lucens* sensu K. Schum. in Engl., P.O.A. C: 93 (1895) sub *lucers* (orth. error); A.Benn. in F.T.A. 8: 221 (1901); Wright in Johnston, Uganda Protect. 1: 348 (1902); Graebner in Z.A.E. 2: 42 (1910); A. Peter in Abh. Ges. Wiss. Göttingen 13 (2): 18, 108 (1928) & in F.D.O.-A.: 113 (1929); Jex-Blake, Wild Fl. Kenya: 112 (1948), *non* L.].

[*P. fluitans* sensu A.Benn. in J.B. 33: 138 (1895); ? K. Schum. in Engl., P.O.A. C: 93 (1895); A.Benn. in F.C. 7: 46 (1897); A. Peter in Abh. Ges. Wiss. Göttingen 13 (2): 18, 107 (1928) & in F.D.O.-A.: 112 (1929), *non* Roth.]

[*P. lucens* L. var. *fluitans* sensu A.Benn. in F.T.A. 8: 222 (1901), *non* Coss. & Germ.]

P. capensis [A.Benn. in Ann. K. K. Naturhist. Mus. Wien 7: 287 (1892), *nom. nudum*;] Hagstr. in Kungl. Svensk. Vet.-Akad., Handl., n.s. 55 (5): 203, fig. 101 (1916). Type: South Africa, Eastern Cape, Uitenhage District: Zwartkopsrivier, *Zeyher* 919 (S, lecto., BREM, LD, isolecto.)

NOTE. From the beginning *P. schweinfurthii*, which is a very variable species, has been confused with other broad-leaved species. When Bennett published this name [in F.T.A. 8: 220 (1901)], he cited three syntypes, of which two, including one later chosen as lectotype by Dandy [in J.L.S. 50: 526 (1937)], proved to belong to *P. nodosus*. However, the third of these collections (*Schimper* 1359) corresponds to the species largely known under the former name which has consequently been accepted for conservation [see Kaplan & Symoens in Taxon, 53: 837–838 (2004) & in J.L.S. 148: 346 (2005)]. *P. schweinfurthii* differs from *P. nodosus* by its sessile or very shortly petiolate submerged leaves and particularly by the characteristics of its stem anatomy.

The plants with only submerged leaves closely resemble *P. lucens*, which explains why they have long been identified under this name. *P. lucens* may be distinguished by its more elliptical leaves, 25–65 mm broad and mostly 3–6 times as long as wide, with often embossed surface, broadly acute to rounded and always distinctly mucronate apex. However some plants (e.g. *Lye* 5245 from Lake Mutanda, Uganda) are really intermediate between the two, so that it is difficult to assign them to one or the other species.

Denny & Lye suggested the existence of two heterophyllous taxa: one fertile, *P. schweinfurthii* and one sterile, being the putative hybrid *P. schweinfurthii* × *P. richardii* (*P. thunbergii* sensu Obermeyer) which they described as *Potamogeton* × *bunyonyiensis* Denny & Lye [in Kew Bull. 28: 120, fig. 1/l–o (1973)], based on a collection from Uganda (*Lye* 5216, K, holotype; MHU, isotype). According to these authors, this hybrid could be distinguished from *P. schweinfurthii* and *P. richardii* by being virtually sterile and having some characters of each parent: submerged leaves, like *P. schweinfurthii*, and coriaceous leaves being predominant, although more lanceolate than those of *P. richardii*. More recently, Lye (in Fl. Somal. 4: 14. 1995, and in Fl. Eth. & Eritr. 6: 23. 1997) considered *P. schweinfurthii* itself to be of hybrid origin, one of its parent being *P. lucens* and the other *P. richardii* (*P. thunbergii* sensu Obermeyer) or *P. nodosus*. Molecular studies and more transplantation experiments are necessary to establish the relationships of these taxa.

Sometimes invasive and considered as a pest in furrows and reservoirs. Consumed as food by Lymnaeid and Planorbid snails.

6. **Potamogeton nodosus** *Poir.* in Lam., Encycl. Méthod. Bot., Suppl. 4: 535 (1816); Dandy in J.L.S. 50: 531, map fig. 7 (1937) pro parte; Cook, Aquat. Wetl. Pl. S. Afr.: 235 (2004); Kaplan & Symoens in J.L.S. 148: 332, fig. 1, 2, fig. 3 map (2005). Type: Canary Is., *Broussonet* s.n. (P!, lecto., FI-W!, isolecto.)

Aquatic herb with submerged and floating leaves; rhizomes robust, long-creeping, biennial or perennial, much-branched, producing apical fusiform turions; stem short-lived or annual, slender to robust, terete, to 2.5 m long, unbranched or sparingly branched. Submerged leaves pale green when fresh and green or brownish green when dried, petiolate, lanceolate to oblong-lanceolate, often absent after fructification; petiole (20–)30–150 mm long; lamina 50–280 mm long, 10–38(–50) mm wide, translucent, 11–21-veined, finely denticulate, at least when young, gradually tapering to a cuneate base and an acute apex; floating leaves bright green, petiolate; petiole (30–)100–210 mm, mostly longer than lamina, ± brown-reddish; lamina lanceolate-elliptical or oblong-elliptical to broadly elliptical or obovate, (40–)50–150 mm long, 20–50(–60) mm wide, 11–23-veined, opaque but sometimes only slightly leathered, cuneate or ± rounded at base, not discoloured and without folds at its base, apex acute to slightly obtuse; stipules axillary, open, convolute and enfolding the stem, 20–60(–125) mm, with 2 veins more prominent than the others and forming dorsal ridges. Peduncles 40–130 mm, not or slightly thickened above; spikes cylindrical, 14–50 mm long, 4–10 mm wide, with up to 15 contiguous flower whorls. Flowers numerous, tepals 0.2–0.3 mm long, suborbicular, carpels (3–)4. Fruitlets brown or reddish brown, 2.7–4 mm long, ventrally nearly straight, dorsal keel sharp, the lateral rather prominent, beak 0.3–0.5(–0.8) mm long, straight or slightly recurved. Stem anatomy: stele of trio or proto type, endodermis of O-type, interlacunar bundles absent, subepidemal bundles absent, peudohypodermis absent.

TANZANIA. Kigoma/Buha Districts: Malagarasi Swamps, Katale [Katare], 29 Aug. 1952, *Lowe* s.n.!; Songea District: Likondi [Likonde] R., 26 June 1956, *Milne-Redhead & Taylor* 10908!

DISTR. **T** 4, 8; widely distributed in Africa; North Africa (from Morocco to Egypt, including Saharan oases); also in Macaronesia, Socotra, Seychelles and Mascarene Islands, Europe, southwestern, central and tropical Asia to Indonesia, Philippines and New Guinea, North and central America

HAB. Shallow parts of lakes, ponds, pans, pools, swamps, also in rivers, often in muddy soil; water depth 1.5–2 m; sometimes in extensive colonies, associated with *Nymphaea caerulea*, *N. indica*, etc.; ± 750 m

SYN. [*P. natans* sensu Thunb., Prodr.: 32 (1794), *non* L.]

P. thunbergii Cham. & Schltdl. in Linnaea 2: 221, t. 6, fig. 21 (1827); Obermeyer in F.S.A., 1: 67, fig. 19 (1966) pro parte. Type: South Africa, Western Cape Province, Hartebeestkraal, near Brak R., Jan. 1819, *Mundt & Maire* (B†, holo.; HAL, lecto.; LE, isolecto.)

P. stagnorum Hagstr. in Kungl. Svensk. Vet.-Akad. Handl., n.s. 55 (5): 159, fig. 78 (1916) & in R.E. Fries, Wiss. Ergebn. Schwed. Rhod.-Kongo-Exped.: 187, fig. 16 (1916); R.E. Fries, Wiss. Ergebn. Schwed. Rhod.-Kongo-Exped., Ergänzungsheft: 61, 72 (1921). Type: Zambia, Chimona R. at Lake Bangweulu, 20 Sept.1911, *Fries* 691 (UPS, holo., LD, iso.)

Fig. 4. *POTAMOGETON RICHARDII* — **1**, habit, × $1/_2$; **2**, flowering spike, × 2; **3**, flower, × 10; **4**, flower, L.S., × 10; **5**, stamen and tepal, × 10; **6**, fruiting spike, × 1; **7**, fruitlet, × 4; **8**, seed, × 8. From *Lebrun* 7909. Drawn by A. Cleuter, and reproduced from F.A.C. with permission.

NOTE. Although widespread from the Maghreb countries to the Cape Provinces of South Africa, *P. nodosus* has long been overlooked on the African continent. Due to the great morphological variation in *Potamogeton* taxa, it is sometimes difficult to distinguish *P. nodosus* from two other broad-leaved species, *P. schweinfurthii* and *P. richardii*. Moreover, published inadequate synonymies brought much confusion between these species. *P. nodosus* may be mostly recognised by its long-petiolate, well developed submerged leaves with a narrowly obtuse to subacute but never sharply mucronate apex. In case of doubt, identification of this species based on stem anatomy is generally successful. If the endodermis is of the O type and there are no interlacunar bundles in the cortex (or rarely one or few), the specimen belongs to *P. nodosus*. On the contrary, complete sessile or subsessile submerged leaves, or endodermis with U-cells and at least one circle of interlacunar bundles exclude *P. nodosus*.

7. **Potamogeton richardii** *Solms* in Schweinf., Beitr. Fl. Aethiop. 1: 194, *in obs.* (1867); A. Benn. in F.T.A. 9: 219 (1901); Graebner in Pflanzenr. 4 (11): 56 (1907); Chiov., Racc. Bot. Miss. Consol. Kenya: 124 (1935); Robyns & Tournay, F.P.N.A. 3: 16, t. 1 (1955); Lind & Tallantire, Flow. Pl. Uganda, ed. 1: 194, fig. 126 (1962) & ed. 2: 194 (1972); Heriz-Smith, Wild Fl. Nairobi R. Nat. Park: 54 (1962); U.K.W.F. ed. 1: 652 (1974); Kaplan & Symoens in J.L.S. 148: 340, fig. 4, 5, fig. 6 map (2005). Type: Ethiopia, Tigray, near Adwa, *Schimper* 135 (K!, lecto.; BM!, FI-W!, G!, K!, L!, LG!, MPU!, P!, U!, isolecto.) and Shire, *Quartin-Dillon* 86 (P!, syn.)

Aquatic herb; rhizomes slender to robust, terete, pink to orange brown, perennial, abundantly branched, with internodes up to 8 cm long; with apical scaly turions; stem terete, up to 1 m long, mostly unbranched, rooting along the nodes, annual. Submerged leaves decaying early, generally not present on adult plants, rarely 1 or 2 at the fruiting stage; petiole (10–)45–180 mm long, 0.2–1.8 times as long as the lamina; lamina bright green to dark green, lanceolate to oblong, sometimes almost reduced to phyllodes, 80–200 mm long, 5–27 mm wide, 5–13 times as long as wide, 5–9(–15)-veined, with narrow rows of lacunae bordering the midrib, entire at margins, narrowly cuneate at base, gradually narrowed towards an obtuse apex, never mucronate. Intermediate leaves often present; floating leaves always present on adult plants; petiole 18–110(–200) mm long, (0.3–)0.6–3 times as long as the lamina, often with a discoloured section at the junction with the lamina, though sometimes only on some leaves or not apparent; lamina green, brownish green to dark green, or ± pinkish, elliptical to oblong-ovate, 30–80(–124) mm long, (11–)20–40(–48) mm wide, 1.6–3.5(–5) times as long as wide, opaque, coriaceous, 11–25-veined, broadly cuneate to rounded at base, broadly acute at apex, entire at margins; stipules axillary, robust, convolute, 25–40(–60) mm long, translucent, persistent as grey fibres after decay. Peduncles inserted in the axils of floating leaves, 50–100 mm long, 2–3.5 times as long as the fruiting spike, as thick as the stem or slightly thinner, a little pale brown tinged; spikes emergent, cylindrical, sometimes crooked, 30–50 mm long in fruit, contiguous. Flowers numerous; tepals pale green; anthers white to yellowish; carpels 4. Fruitlets green to pale brown, (3.2–)3.9–5.2(–5.5) mm long, 2.1–3.3 mm wide, ventrally straight, dorsally convex, distinctly 3-keeled. Stem anatomy: stele of trio type, endodermis of U-type, interlacunar bundles present in (1–)2(–3) circles, subepidermal bundles mostly present, pseudohypodermis present, 1-layered. Fig. 4 (p. 14).

UGANDA. Kigezi District: near Kigezi, Lake Bunyoni, Nov. 1940, *Eggeling* 4237!; Mengo District: Kabaka's Lake, Jan. 1936, *Hancock & Chandler* 135!; Masaka District: near Kaboyo, 12 July 1971, *Lye* 6459!

KENYA. Turkana District: Murua Nysigar [Moruassigar], 18 Feb. 1965, *Newbould* 7277!; SW slopes of Elgon, 30 Dec. 1952, *R. Ross* 1346!; Nairobi District: outskirts of Nairobi, Karura Forest, 21 March 1953, *Verdcourt & Steele* 913!

TANZANIA. Bukoba District: Kishanda, Dec. 1931, *Haarer* 2110!; Arusha District: Longil Swamp, 3 Apr. 1968, *Greenway & Kanuri* 14012!; Iringa District: Mufindi, Lake Luisenga, 15 March 1962, *Polhill & Paulo* 1767!

DISTR. **U** 2, 4; **K** 2–6; **T** 1, 2, 4, 5, 7, 8; also in Cameroon, Ethiopia, Eritrea, Somalia, Congo (Kinshasa), Rwanda, Burundi, ? Angola, Zambia, Malawi, Zimbabwe, Namibia, Lesotho, South Africa, Madagascar.

HAB. Lakes, ponds, pools, also in streams, often in acidic water, water depth 30–100 cm; 1150–3450 m

SYN. *P. americanus* Cham. & Schltd. var. *richardii* (Solms) Schweinf. in Bull. Herb. Boissier 2 (Append. 2): 8 (1894)

[*P. natans* sensu A.Benn. in F.C. 7: 46 (1897); Graebner in Z.A.E. 2: 41 (1910); A. Peter in Abh. Ges. Wiss. Göttingen 13: 108 (1928) & F.D.O.-A.: 113 (1929), *non* L.]

[?*P. coloratus* sensu A.Benn. in F.T.A. 8: 222 (1901); Graebner in Pflanzenr. 4, 11: 69 (1907) pro parte; A. Peter in Abh. Ges. Wiss.Göttingen 13: 107 (1928) & F.D.O.-A.: 113 (1929), *non* Vahl ex Hornem.]

[*P. nodosus* sensu Dandy in J.L.S. 50: 531, map fig. 7 (1937) pro parte; Friis *et al.* in Op. Bot. 63: 55 (1982), *non* Poir. in Lam.]

[*P. thunbergii* sensu Obermeyer in F.S.A. 1: 67, fig. 19 (1966); F.P.U., ed. 2: 194, fig. 126 (1972); Lisowski *et al.* in F.A.C., Potamogetonaceae: 9, t. 2 (1978); Symoens *et al.* in B.S.R.B.B. 112: 79 & ss., fig. 3, map 3 (1979); U.K.W.F., ed. 2: 302 (1994); Lye in Fl. Eth. 6: 23, fig. 176.1.6 & 7 (1997); Wiegleb & Kaplan in Folia Geobot. 33: 264 (1998); Cook, Aquat. Wetl. Pl. S. Afr.: 238, fig. 256 a, b (2004), *non* Cham. & Schltd. (1894)]

NOTE. *P. richardii* has often been confused with *P. nodosus.* It may be distinguished from this species by its submerged leaves promptly decaying, by its floating leaves with often discoloured petiole at the junction with the lamina and lamina mostly rounded at base, by its mature spikes more densely fructified and by its larger fruits (3.2–5.5 × 2.1–3.3 mm in *P. richardii*, 2.7–4 × 1.7–2.7 mm in *P. nodosus*). The best diagnostic characters reside in the stem anatomy: *P. richardii* has endodermis cells with U-thickenings, interlacunar bundles are present; in *P. nodosus*, the endodermis cells have O-thickenings and there are no interlacunar bundles.

INDEX TO POTAMOGETONACEAE

No new names validated in this part

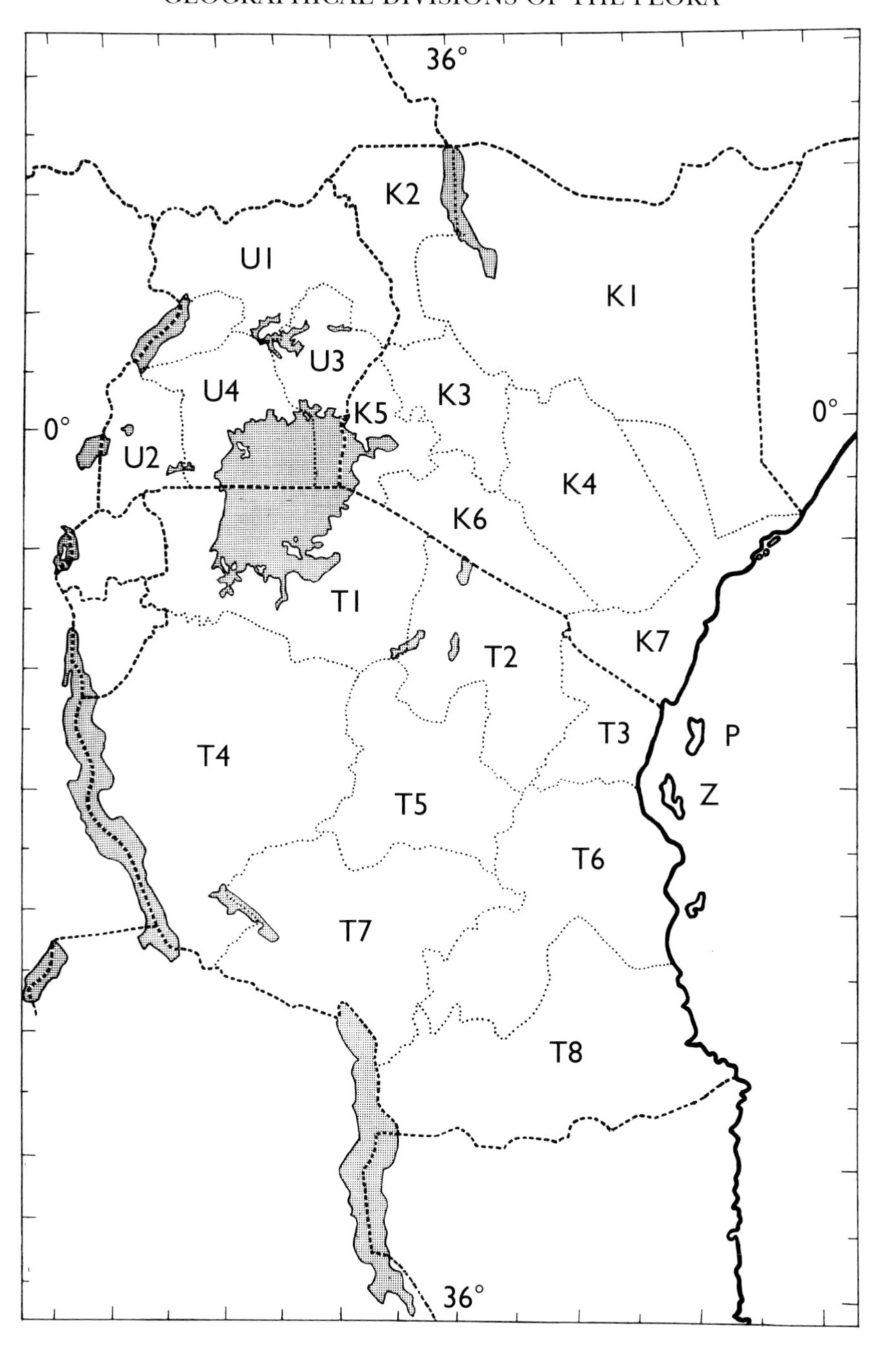
36°
K2
U1
K1
U3
U4
K3
K5
0°
0°
U2
K4
K6
T1
K7
T2
T3
P
T4
Z
T5
T6
T7
T8
36°

First published in 2006 by
Royal Botanic Gardens, Kew
Richmond, Surrey, TW9 3AB, UK
www.kew.org

ISBN 1 84246 130 3

Design by Media Resources, typesetting and page layout by Margaret Newman,
Information Services Department,
Royal Botanic Gardens, Kew.

Printed in the UK by Hobbs the Printers

For information or to purchase all Kew titles please visit
www.kewbooks.com or email publishing@kew.org